GENETIC *Diversity*

TEACHER SUPPLEMENT

GENETIC *Diversity*

TEACHER SUPPLEMENT

INSTITUTE *for* CREATION RESEARCH

Science Education Essentials is a curriculum supplement series designed to cover vital topics in the various science disciplines, all from a thoroughly biblical viewpoint. Each product includes a teacher instructional guide, along with K-12 activities and classroom helps to guide discussion, reinforce subject content, and facilitate hands-on laboratory exercises.

Published by the Institute for Creation Research.

Science Education Essentials

Series Creator: Dr. Patricia L. Nason

Project Manager: Janis McCombs

Managing Editor: Beth Mull

Assistant Editor: Christine Dao

Graphic Designer: Susan Windsor

Science Reviewers: Dr. John Morris, Dr. Charles McCombs, Dr. Jeff Tomkins, Dr. Randy Guliuzza, Dr. Chris Osborne, Dr. Larry Vardiman, Dr. Tim Clarey, Dr. Brad Forlow, Brian Thomas, Frank Sherwin

ISBN: 978-0-932766-96-0

Genetic Diversity

Teacher Supplement Authors:

Dr. Gary Parker graduated magna cum laude in biology/chemistry from Wabash College in 1962, where he was also elected to the national scholastic honorary society Phi Beta Kappa. His master's thesis was published in *Copeia* and presented at a major scientific conference. His doctoral work, supported by a competitive 15-month fellowship grant from the National Science Foundation, was completed as he emerged as a creationist, enabling him to add paleontology as an emphasis to his Ed.D. in Biology/Geology from Ball State University in 1973. Known for his warmth and casual style, Dr. Gary Parker is a creation science speaker and writer popular across the United States and around the world. He is also the author of numerous books, including *Creation: Facts of Life* and *Building Blocks in Science.*

Dr. Jeffrey Tomkins earned a Master's in Plant Science in 1990 from the University of Idaho, where he performed research in plant hormones. He received his Ph.D. in Genetics from Clemson University in 1996. While at Clemson, he worked as a research technician in a plant breeding/genetics program, with a research focus in the area of quantitative and physiological genetics in soybean. After receiving his Ph.D., he worked at a genomics institute and became a faculty member in the Department of Genetics and Biochemistry at Clemson. He had become a Christian while an undergraduate at Washington State University in 1982, with a goal to eventually work as a scientist and author in the creation science field. In 2009, Dr. Tomkins joined the Institute for Creation Research as Research Associate.

K-12 Instructional Contributors: Dr. Gary Parker, Dr. Patricia Nason, Dr. Charles McCombs, Janis McCombs, Leona Criswell, Frank Sherwin

Dr. Parker has taught similar content in his classic book *Creation Facts of Life* (Master Books, 2006).

For additional resources from the Institute for Creation Research, please visit www.icr.org or call 800.337.0375.

ISBN: 978-1-935587-04-0

Printed in the United States of America.

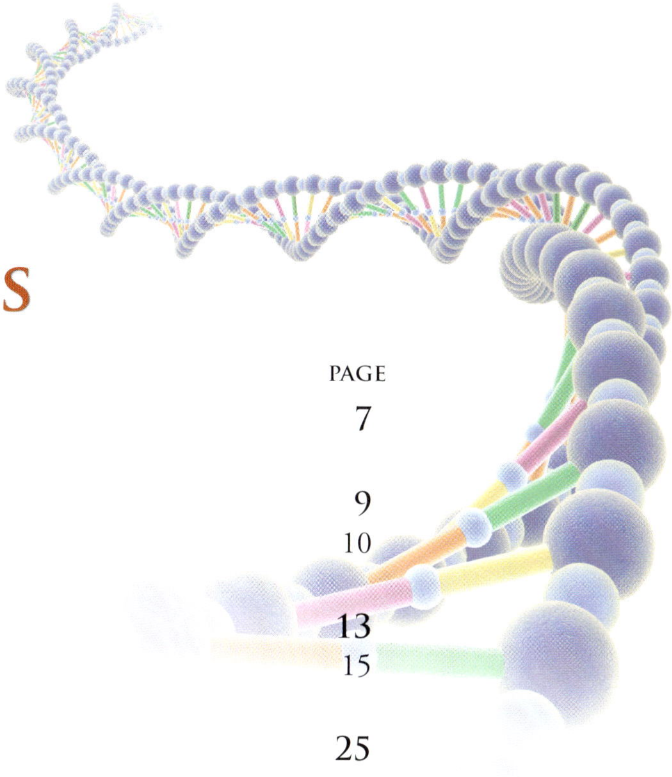

TABLE OF CONTENTS

	PAGE
PREFACE	7
VARIETY: THE SPICE OF LIFE	9
Scientific Classification	10
VARIATION: SPECIES AND KIND	13
Genetic Basis for Variation	15
DIVERSITY AND THE MOSAIC CONCEPT	25
CONCLUSION	39
BIBLIOGRAPHY	41

PREFACE

Teachers mold the minds of their students, helping them construct knowledge and an understanding of the world around them. A teacher's influence on the belief system, as well as cognitions, of a student can affect the student for a lifetime.

For 40 years, the Institute for Creation Research has equipped teachers with evidence of the accuracy and authority of Scripture. In keeping with this mission, ICR presents Science Education Essentials, a series of science teaching supplements that exemplifies what ICR does best—providing solid answers for the tough questions teachers face about science and origins.

This series promotes a biblical worldview by presenting conceptual knowledge and comprehension of the science that supports creation. The supplements help teachers approach the content and Bible with ease and with the authority needed to help their students build a defense for Genesis 1-11.

Each science teaching supplement includes:

- A content book written at the high school level to give teachers the background knowledge necessary to teach the concepts of scientific creationism with confidence. Each content book is written and reviewed by creation scientists, and can be purchased separately in class sets.
- A CD-ROM packed with teacher resources, including K-12 reproducible activities and PowerPoint presentations. The instructional materials have been pilot tested for ease in following instructions and completeness of activities. They have also been reviewed by scientists for scientific accuracy and by theologians for biblical correctness.

Science Education Essentials are designed to work within a school's existing curriculum, with an uncompromising foundation of creation-based science instruction. Secular textbooks are finding their way into Christian schools. Teachers may not lack belief in the Word of God, but they often do not have adequate information or knowledge concerning the tenets of scientific and/or biblical creation. Science Education Essentials equips teachers with the tools they need to teach the science of origins from a biblical rather than an evolutionary worldview.

The goal of each science supplement is to:

a) increase the teacher's understanding of and confidence in scientific creation and the truth of God's Word, while glorifying God as Creator;

   But sanctify the Lord God in your hearts: and be ready always to give an answer to every man that asketh you a reason of the hope that is in you with meekness and fear. (1 Peter 3:15)

b) provide teachers with a toolkit of activities and other instructional materials that build a foundation for their students in creation science apologetics;

Beware lest any man spoil you through philosophy and vain deceit, after the tradition of men, after the rudiments of the world, and not after Christ. (Colossians 2:8)

c) encourage the use of the higher level thinking necessary to stand firm against the lies of evolution and humanism.

...that we henceforth be no more children, tossed to and fro, and carried about with every wind of doctrine, by the sleight [trickery] of men, and cunning craftiness, whereby they lie in wait to deceive; but speaking the truth in love, may grow up into him in all things, which is the head, even Christ. (Ephesians 4:14-15)

With Science Education Essentials, teachers can equip the future generation of scientists and individuals to examine the evidence for the truth of Scriptures through an understanding of creation science. By using hands-on activities and relating scientific truth to the Bible, teachers/parents will be grounding their children in creation science truths so they can provide a logical response when challenged with science that is based on a philosophy that is in direct contradiction to Genesis 1-11.

As the leading creation science research organization, ICR is providing meaningful creation science material for classroom use. Our desire is that the materials renew minds, defend truth, and transform culture (Romans 12:1-2) for the glory of the Creator.

Dr. Patricia L. Nason

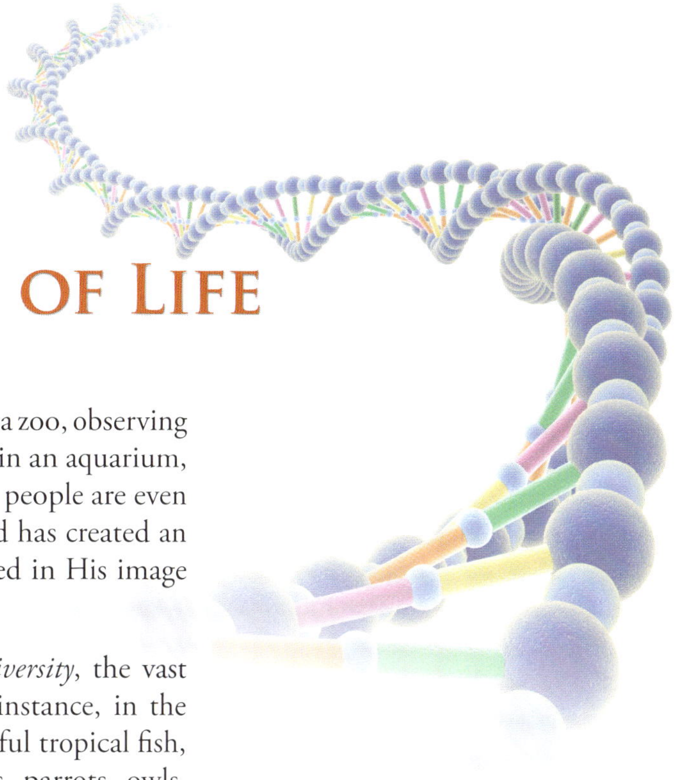

VARIETY: THE SPICE OF LIFE

Almost everyone enjoys seeing the wonderful animals at a zoo, observing wild animals in a national park, gazing at undersea life in an aquarium, or watching birds gathered around a bird feeder. Some people are even fascinated by creatures such as snakes and insects. God has created an incredible variety of creatures and we have been created in His image to enjoy this variety as the "spice of life."

Variety actually exists on two levels. First, there is *diversity*, the vast number of dramatically different created kinds. For instance, in the sea we find microscopic plankton, spiny lobsters, colorful tropical fish, squid, dolphins, and whales that weigh a ton. Doves, parrots, owls, eagles, and vultures soar gracefully through the skies. We also find spiders, lizards, mice, dogs, cats, cows, and in some areas even panthers, antelope, and bears inhabiting the lands around us.

Second, there is *variation*, differences within the created kinds. Think about the many different types of dogs you can find in a pet store, including Chihuahuas, terriers, schnauzers, retrievers, hounds, spaniels, bulldogs, and poodles. Have you ever ridden a horse, watched children ride ponies at a county fair, seen a 17-inch miniature horse, or watched mighty Clydesdales work the land? They have a wide variety of sizes and strength, but they are all equines. Whether you are at a mall in a large city or at a local diner in a small town, look at the individuals around you and consider the tremendous variation that occurs among the people you see on a daily basis.

Most people are content to simply enjoy the vast variety both within and among living kinds (variation and diversity). But scientists want to classify the differences. Why? There are many reasons:

1. God created the human mind to recognize patterns. Categorizing diverse data into groups (e.g., trees, flowers, birds, dogs, etc.) helps people to organize, retrieve, and see patterns among otherwise overwhelming bits of information.

2. Classifying living creatures helps scientists apply what is learned about one organism to others with similar structures and/or functions.

3. There is even a spiritual dimension to the classification of living things. The current form of the crucial and continuing battle between creation and evolution began with a battle over the "origin of species," the extent of variation within a kind, and whether diversity among kinds reflects common ancestry (evolution) or theme and variation in the Creator's plan.

Scientific Classification

Carolus Linnaeus

The science of classification is called biosystematics. The system that scientists use today for identifying and grouping distinctive kinds of organisms was developed in the 1700s by Swedish biologist and Christian creationist Carolus Linnaeus. (This is the Latin form of his given Swedish name, Carl von Linné.) Latin and Greek were the cross-cultural languages of intellectuals in Linnaeus' time and he assigned scientific names to plants and animals using Latin and Greek roots, prefixes, and suffixes. His system worked so well that in the mid to late 1800s Darwin and Darwin's followers (evolutionists) continued to use the creationist's system. However, they tried to change the meaning of key words to suggest that continuous change through millions of years of struggle and death (Darwin's "war of nature") produced each kind, as opposed to the plan, purpose, and special acts of God's creation at the beginning of earth's history.

What is a gopher? For some people, it is a furry animal that digs holes in the yard. For others, it refers to a type of tortoise, and still others call a snake by that name. For a few, it is the kind of wood that Noah used to build the Ark. To bring scientific order out of the chaos of common names, Linnaeus used binomial nomenclature, a "two-name naming system," to give each different kind of organism a unique scientific name that consisted of a genus (general) name and a species (specific) name.

The more general name, the genus name, is written first with the first letter capitalized. The more specific name, the species name, is written second in lower case. The whole, two-part scientific name is usually italicized or otherwise set off from the surrounding type. Committees of scientists working in systematics meet regularly to accept or reject proposals for new scientific names, and they work to ensure that no two different kinds of organisms have the same genus-species name.

The familiar scientific name for mankind is *Homo sapiens,* which means in Latin (somewhat optimistically) "man of wisdom." The scientific name for a white oak is *Quercus alba* (oak is the general name and white is the specific), and the red oak (or Oak red) is *Quercus rubrum.* The species name *rubrum* (or *rufus*) is used in various scientific names—for example, *Acer rubrum* (red maple) and *Macropus rufus* (red kangaroo). After the initial appearance of a genus name in a text, it can be abbreviated. *Quercus alba* can be written *Q. alba* in later references. The plural of genus is genera. The plural of species is also species. ("Specie" is an error, not the singular of species.)

The Linnaean system of classification also groups objectively identifiable kinds into progressively higher taxonomic ranks, from family to order, class, phylum or division, and kingdom, with domain recently added at the "top." Taxonomists use observations of precisely defined characteristics to give an organism its proper scientific name. Organisms are grouped on the basis of shared characteristics, with "lower" ranks sharing many small features. Organisms grouped into "higher" categories share a few traits that are considered more important. Assignment to a higher taxonomic group is much more subjective than assignment of the scientific genus-species name, and potentially much more controversial. Still, many names for families, orders, and classes enjoy wide usage among taxonomists.

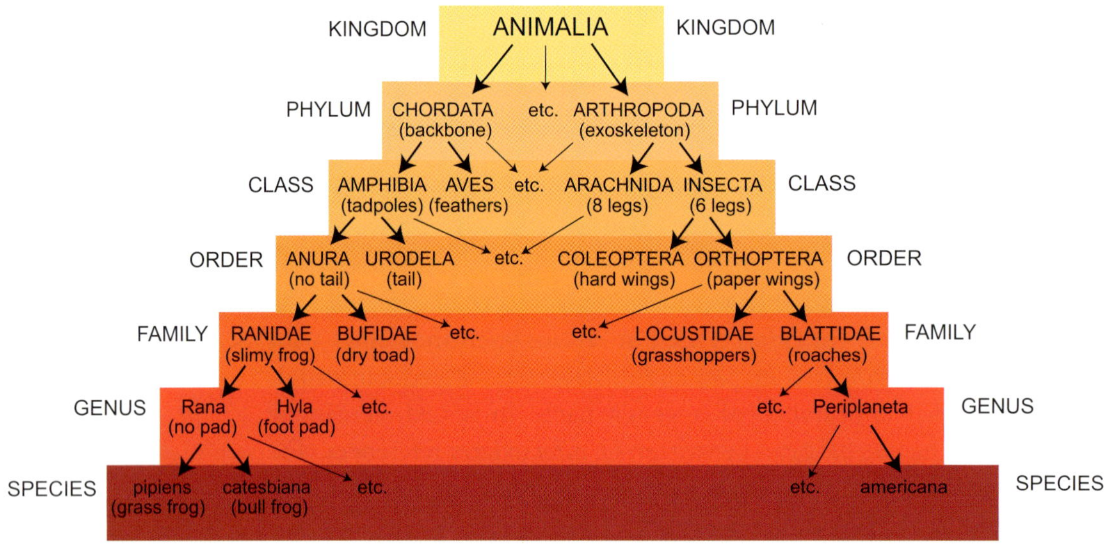

Taxonomy is the practical "how" of classification; systematics is the theoretical "why." Systematics obviously deals with more subjective and controversial questions, including:

1. What criteria should be used in classification and at which level or rank?

2. Which characteristics are more important and why?

3. What range of trait variation allows organisms to be given the same scientific name?

4. What limits to variation within kind can be scientifically identified, if any?

5. Finally, which worldview—creationist or evolutionist—makes the most sense of scientific data regarding life's variety, both variation within kind and diversity among kinds?

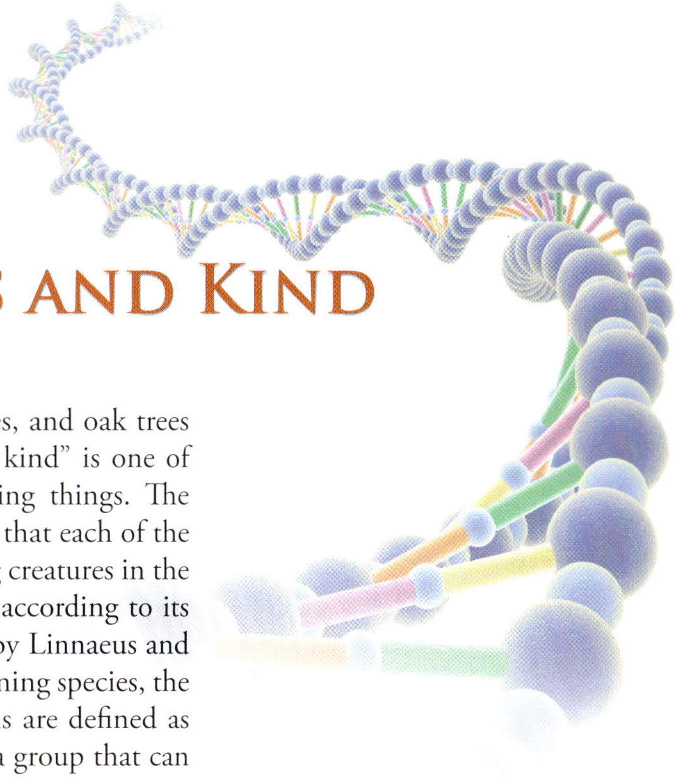

Variation: Species and Kind

Dogs have puppies, cats have kittens, cows have calves, and oak trees bear acorns that grow more oaks. "Multiplying after kind" is one of the most distinctive and best-known features of living things. The first chapter of Genesis repeatedly and explicitly states that each of the many different kinds of life—plants, trees, birds, living creatures in the water and on land—was created by God to multiply according to its kind. This ability to reproduce (interbreed) was used by Linnaeus and is still used by scientists today as the chief basis for defining species, the scientific relative of the biblical term *kind*. Organisms are defined as belonging to the same species if they are members of a group that can produce fertile offspring.

All scientists, even those who believe in Darwinian evolution, still honor "after kind" (interbreeding) as the primary and most objective definition of species. While paying lip service to this scientifically verifiable classification of species, however, some of Darwin's followers are increasingly and flagrantly violating this objective (and creationist) definition of species in favor of subjective opinions about trivial traits, and are thus outrageously multiplying the number of species names and turning the science of systematics into a morass of subjectivity.

As a pigeon breeder, Darwin himself was well aware of the variation within kind. An incredible variety of pigeons can be bred from, and back to, the pigeon commonly found around city parks, statues, and farms, including pigeons with a fancy spray of tail feathers, pigeons with expandable neck pouches, skinny and plump pigeons, and pigeons with dull or dazzling colors. In a university debate, an evolutionist opponent even said that bone structures among Darwin's pigeons were so different that their fossils would be classified as different species—except, he continued, we know they are all one species because all can still interbreed.

Trivializing species names has even endangered "endangered species"! Too often a species name, and species protection, has been given to a very small subset of a larger interbreeding population, the true (or scientifically objective) species. Take, for example, the case of the endangered Florida panther. What endangered it? Hunters? Alligators? Global warming? Automobiles? No, it was endangered by mutations (which are often called "the raw materials for evolutionary progress"). These poor panthers were so inbred and riddled with mutational defects in genes that affected their circulatory and reproductive systems that they could not produce cubs that would survive to reproductive maturity. Had the Florida panther been "protected" as a true species, it would now be completing its journey to extinction.

Fortunately, park rangers more interested in panthers than in politically correct (and scientifically incorrect) semantics brought in panthers from the western United States to crossbreed with the Florida panthers. Since the Florida panthers were not really a different species at all but only a variation of the same created kind or true species, they freely interbred. Both varieties of panthers had defective mutants among their genes—but the western and Florida panthers had different defects. Many defects do not show up in offspring unless both parents have the same defect, a situation that is far less likely in outbreeding than with inbreeding. Panthers are now making a comeback in Florida. They are no longer "pure" Florida panthers, just healthier panthers! Other "endangered species" that are kept imprisoned with inbreeding and defective genes would benefit greatly from access to all the variation God built into each created kind!

You have probably heard the phrase "extinction is forever." That is true for a real created kind or true species. Each kind was created with a special role to play in the total web of life, so we should support efforts to preserve created kinds, both to honor the God who created them and to maintain the earth's ecological health. But extinction of varieties within a kind, or "subspecies," is *not* forever. The variety within a kind (species) can be bred back by crosses with others of its created kind or true species.

Take the quagga, for example, a "horse" with faint stripes like those of a zebra on its forequarters. Quaggas were once thought to have become extinct when the last one died in the Artis Magistra Zoo in Amsterdam in 1883. But breeders are now bringing them back by crossing horses and zebras. Horses and zebras have each been given several different species names—in direct violation of the most objective and scientifically testable definition of species. Since they actually are only variations in created "horse kind," horses and zebras can and do interbreed and the "resurrected" quagga is only one result. The "zorse" in the Creation Museum near Cincinnati is another.

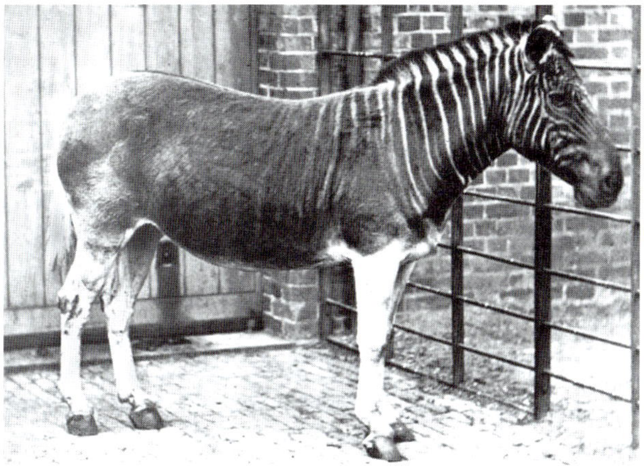

Quagga from the London Zoo, 1870

Genetic Basis for Variation

In the years since Linnaeus, scientists have discovered an increasing number of factors that limit reproduction to "after kind." For example, sperm and egg cells have recognition factors on their surfaces (such as the proteins fertilizin and anti-fertilizin) that enable only the sperm and egg of the same kind to unite. Sperm-egg recognition pairing is vital to the many sea creatures that shed their egg and sperm into the ocean and just let "nature take its course"—nature operating according to God's plan for reproduction after kind!

Organisms of the same kind have genes of the same kind. For any given gene, the actual DNA sequence may vary between individuals by one or a few base pairs, resulting in variant forms of the gene called *alleles*. For example, there are over 300 alleles of the human adult hemoglobin gene. That is a lot of variation, but all those alleles produce hemoglobin, a protein for carrying oxygen in red blood cells, so the allelic variation is clearly variation within kind.

Allele: any of the alternative forms of a gene that may occur at a given locus[1]

Genes of the same kind (alleles) can be defined objectively as segments of DNA that occupy corresponding positions (loci; singular, locus) on homologous chromosomes. *Homologous chromosomes* are pairs that usually look alike (except for sex chromosomes), but homologs come from the different parents (one from each), so their genetic content is similar but not identical. Homologs and the alleles they carry pair up and then separate in the kind of cell division (*meiosis*) required for sexual reproductive cycles. During meiotic pairing, alleles may swap places on homologous chromosomes. This process (called *crossing over)* greatly increases variation among offspring of the same parents by adding some of the mother's (maternal) genes to the chromosome that originally

1. Allele. *Merriam-Webster Online Dictionary.* 2010. Merriam-Webster Online. Posted on merriam-webster.com/dictionary.

Darwin's pigeon drawings, from *The Variation of Plants and Animals Under Domestication*

came from the father (paternal) and vice versa, mixing maternal and paternal genes on still-matching homologous chromosome pairs.

Recombination: the formation by the processes of crossing over and independent assortment of new combinations of genes in progeny that did not occur in the parents[2]

Alleles are also turned on and off by the same regulators, which allows for considerable variation in a gene's structural or functional expression (*phenotype*)—but, again, this is only variation within kind. Note that it is not subjective human opinion but objective, scientifically observable cellular processes that tell us which genes are the same kind.

According to Darwin's followers, mutations produce "new genes" that provide the raw material for evolutionary progress. However, this is not the case. All known mutations are only alleles of previously existing genes, and the most change they could ever produce—*no matter what length of time is involved*—is variation within kind. Any hope that bacteria could ever evolve through multiple stages into people would

2. Recombination. *Merriam-Webster Online Dictionary.* 2010. Merriam-Webster Online. Posted on merriam-webster.com/dictionary.

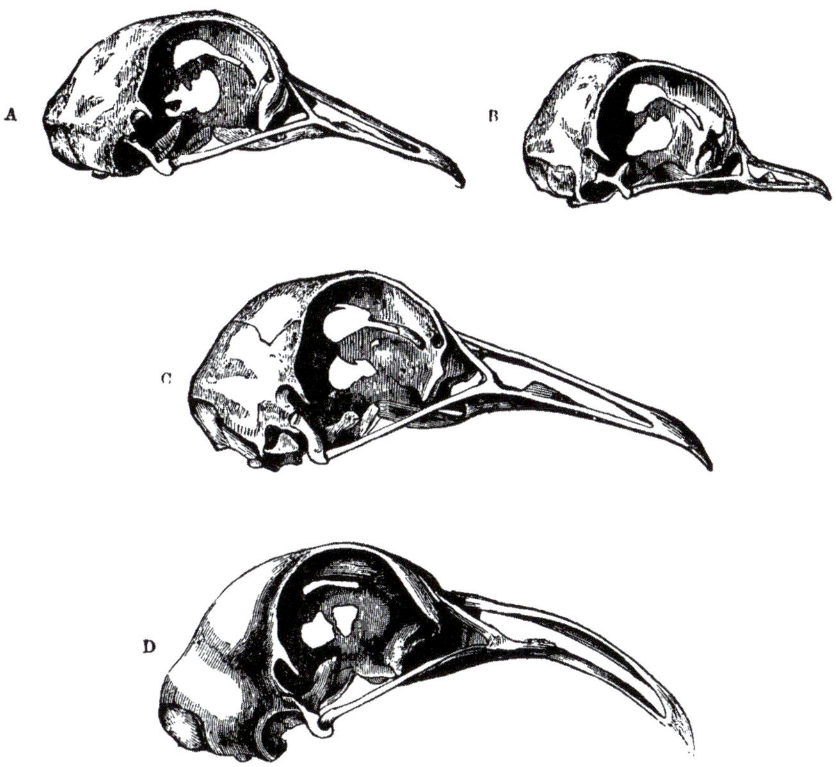

Fig. 24.—Skulls of Pigeons viewed laterally, of natural size. A. Wild Rock-pigeon, *Columba livia*. B. Short-faced Tumbler. C. English Carrier. D. Bagadotten Carrier.

Darwin's pigeon skull drawings, from *The Variation of Plants and Animals Under Domestication*

require something much more than mutational variation. Namely, it would require some process that could produce not just alleles, but wholly new and different genes, including 1) genes with novel information that creates new and different categories of traits; and 2) non-allelic genes that do *not* occupy the same position on homologs, that do *not* pair and cross over in meiosis, and that are *not* turned on and off by the same regulators. Since man's sin, mutations have been producing alleles that introduce defects, deficiencies, disorders, and disease organisms into God's initially perfect creation. Mutations never could give bacterial descendants the novel information required to develop a rhythmic heartbeat, a flying wing, or a seeing eye.

The "production of higher animals," as Darwin called it, would also require adding to the total amount of genetic information, or genome, that one generation of a kind can pass on to the next. The protein-coding genes in the common bacterium *E. coli*, for example, number about 5,000, while there are 25,000 or more in the genome of each human cell. Furthermore, the genome for each kind includes not only protein-coding DNA sequences, but also the modulating and regulatory sequences once falsely called "junk DNA." Molecular biologists currently behave as though members of a kind all have the same genome (e.g., *the* human genome), but population biologists

and forensic ("CSI") scientists often focus on allelic differences that produce structural and functional (phenotypic) variability—i.e., individual differences among members of the same kind. The total amount of genetic information might be called the vertical extent of a kind's *genome*, or (in older terminology) the depth of its *gene pool.* The allelic variability within a genome might be called its horizontal extent, or the width of its gene pool.

All members of a kind can be thought of as having the same genomic depth, but individuals may have different genomic widths. For example, the genetic diversity (or width) of the greyhound gene pool is very low due to inbreeding. Crossing purebred greyhounds results in more dogs with short hair, thin bodies, long snouts, and fast speed. Crossing two mongrels, however, can produce big dogs and small dogs, dark and light and splotchy colored dogs, dogs with long and short hair, yappy and quiet dogs, mean and affectionate dogs. The genetic diversity (or width) of the mongrel's gene pool is quite large compared to the greyhound's, but the depth of the genomes—the numbers of genes and categories of genetic information—is essentially the same in all breeds of dog kind.

Heterozygous: having the two alleles at corresponding loci on homologous chromosomes different for one or more loci[3]

Earlier we considered the vast variation in both animals and in people that we encounter on a daily basis. Is it reasonable to believe that the tremendous variety we see today was "pre-set" into the first breeding pairs? For example, could the genetic code of Adam and Eve have resulted in the tremendous variation of people that currently exists all around the world? It has been reported that human beings are *heterozygous* for 6.7 percent of their genes on average.[4] In other words, less than 7 percent of their genomes have a dissimilar pair of genes for a given hereditary characteristic. These allelic differences in gene pairs are connected to

3. Heterozygous. *Merriam-Webster Online Dictionary.* 2010. Merriam-Webster Online. Posted on merriam-webster.com/dictionary.

4. Ayala, F. 1978. The Mechanisms of Evolution. *Scientific American*. 239 (3): 56-69.

many recognizable traits, including having free or attached earlobes, the ability or inability to roll the tongue, and having or not having a cleft chin or facial dimples. Can this degree of heterozygosity sufficiently explain the variation in the human population? Although the degree of difference in the genome might appear insignificant, calculations based on just 6.7 percent allelic variation show that a human couple in theory could produce 10^{2017} distinct children. The astronomical variety God has genetically programmed into the gene pool is astonishing. Each one of us is indeed uniquely "fearfully and wonderfully" made by our Creator (Psalm 139:14).

The color variations in England's famous peppered moth, *Biston betularia*, are good examples of genetic diversity resulting from allelic differences and recombination. Somewhat like human beings, the moth can vary in color from very light to very dark. Indeed, the proteins that produce shades of darkness in the skin of humans and the scales of moths are both versions of a protein called *melanin.*

From 1850 to 1950, pollution darkened the tree bark where the moths rested, shifting the population from 98 percent light (containing little melanin) to 98 percent dark (containing much melanin). Darwin's followers claim that this was an example of "evolution going on today." It was not. Not only did the moth *not* change species, it did not even acquire a new trait. What the moths really demonstrated was created variability designed to enable kinds to fill different environments. Color variations from very light (peppered) to very dark *(carboneria)* and many gradations between are traceable through insect collections going back for centuries. Since God created the first of each kind to multiply and fill the earth, He may have created the first *Biston betularia* with a speckled or splotchy distribution of melanin on its wing and with a set of genes (genotype) symbolized by the letters AaBb. In one generation, such moths could produce five color forms, grading from very dark (AABB) to dark (three

	AB	Ab	aB	ab
AB	**ABAB**	**ABAb**	**ABaB**	**ABab**
Ab	**AbAB**	**AbAb**	**AbaB**	**Abab**
aB	**aBAB**	**aBAb**	**aBaB**	**aBab**
ab	**abAB**	**abAb**	**abaB**	**abab**

Punnett square showing variations of multiple alleles.

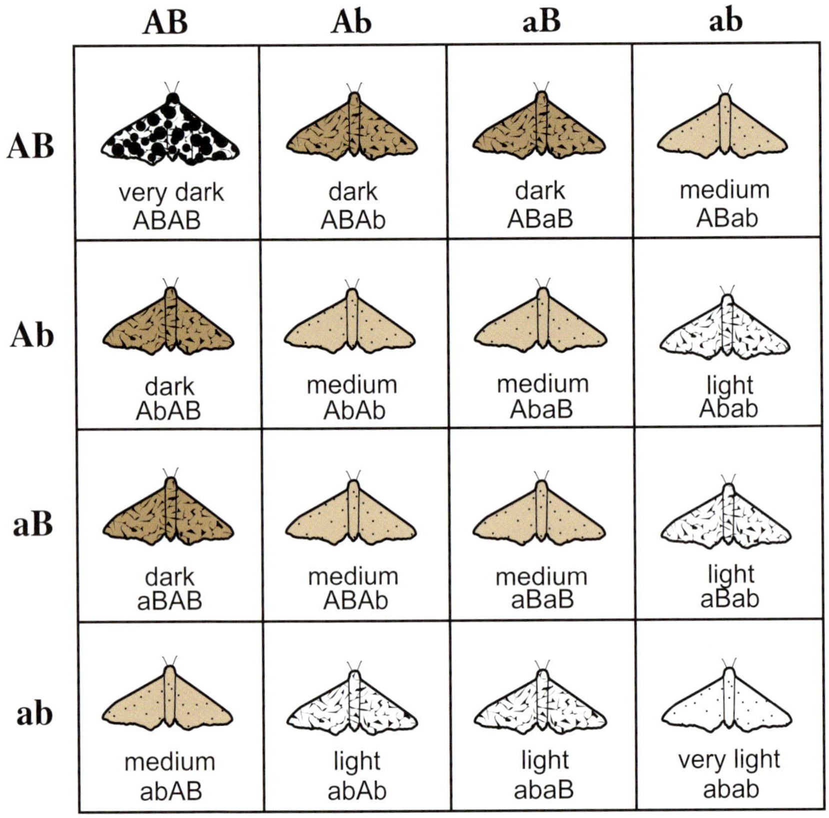

Figure 1: Possible offspring variations from parent moths with different genotypes. The more dominant (capitalized) alleles that are present, the more melanin and darker coloration the offspring will have.

"capital genes"), medium (any two capitals), light (one capital), to very light or peppered (aabb).

There are at least four melanin wing color control alleles at two gene sites (loci) in the moth genome that was just described: A, a, B, b. That total gene pool for wing color control can be found in just one moth with medium wing color (AaBb), or it can be "spread around" among many moths with visibly different wing color phenotypes. (A similar situation applies to skin color among people.) In fact, the gene frequencies (percentages of each gene) in one AaBb "medium-colored" moth are exactly the same as the gene frequencies in the 16 offspring that show five different amounts of wing color, as shown in Figure 1. (Count and see that the Punnett square includes 16 each of A, a, B, and b—each one-fourth of the total, the same as the one-fourth each of A, a, B, and b genes in their parents.) This example shows the tremendous potential for variation due to allelic differences and genetic recombination. All that *individual variation* occurs in a group that remains constant—demonstrating creation and variation within a single created kind!

The creation model predicts that all variation within the kind would have been "pre-set" in the wide gene pool of the original created parents. Due to allelic differences and recombination, a great amount of potential variation in size, color, form, and function would have been present in the genotypes of the original created ancestors of human beings, as well as in the first of each plant and animal kind. The potential genetics for the darkest Nigerian and the lightest Norwegian, the tallest Watusi and the shortest Pygmy, the highest soprano and the lowest bass would have been present right from the beginning in two quite average-looking people.

God designed the original members of each created kind to multiply and fill the earth. The genetic variation "pre-set" within the first breeding pairs (the generalized type of a created kind) would result in offspring with suitable variations to adapt to and fill various environments around the world. As unique and genetically distinct descendants settled in the appropriate environments, they would become geographically isolated from others of their kind. This would result in populations of specialized descendants from the original breeding pair suited for their particular habitat. For instance, the original dog created kind (generalized type) has subsequently produced wolves, coyotes, and foxes, which are specialized descendants. Therefore, due to allelic differences and recombination, the wide genetic pool of general created kinds has produced an array of variation, resulting in numerous subgroups (specialized descendants) that are suitable for and adaptable to particular environments, in accordance with God's mandate to fill the earth. This genetic design results in tremendous variation in those traits that enabled the rapid production of specialized subgroups from general created kinds to fill the earth following both creation and Noah's Flood.

Genetic drift: random changes in gene frequency especially in small populations when leading to preservation or extinction of particular genes[5]

The total amount of genetic information (genome depth) is equivalent for each member of a created kind. The specialized subgroups within a kind have allelic variability (genome width), allelic differences that produce the structural and functional (phenotypic) variability. When subgroups become isolated into small specialized populations, distinctive phenotypic changes can occur due to genetic drift. For example, many members of the Afrikaner population of South Africa are descended from Dutch settlers who had an unusually high frequency of the gene

5. Genetic drift. *Merriam-Webster Online Dictionary.* 2010. Merriam-Webster Online. Posted on merriam-webster.com/dictionary.

that causes Huntington's disease. As a result, their descendants have a higher than average frequency of this gene.

As they spread across the globe, it seems that some descendants of a created kind occasionally became *reproductively isolated* from other varieties of the same created kind. In other words, they lost the ability to interbreed with their original group. Isolated populations that developed from only a small number of individuals (a "genetic bottleneck") have a much smaller gene pool and less genetic diversity than their original group. If species are strictly defined as interbreeding groups, these subgroups of a created kind could be called "new species"—but that is no help to evolutionists. *Speciation, yes; evolution of a new life form, no*. Reproductively isolated groups developing within the same kind have *less* variability than the original kind, which means *less* ability to move into new environments and *less* ability to meet changes in existing environments. "Evolution" based on the break-up of large populations into less variable populations would be evolving toward inbreeding and extinction, not toward Darwin's "production of higher animals"! Evolution from "lower to higher" creatures requires the addition of new types of genes with new information that had never existed before. Instead, the unfolding allelic variation present from the beginning (*entelechy*) displays the richness of variation built into each kind by plan, purpose, and special acts of creation.

Consider 1) the tremendous genetic variability God initially built into each kind, 2) the variability augmented for better or worse by mutation and selection, and 3) the division of some created kinds into reproductively isolated subgroups. All these factors (plus unscientific over-classification and the trivializing of species names) mean that it would be most unwise, despite the initial temptation, to equate the biblical term "kind" with the once scientific, now arbitrary and subjective, term "species."

Using the Hebrew words *bara* (create) and *min* (kind), Frank Marsh, a leading 20th-century creationist biologist, coined the term *baramin* for "created kind." The scientific study of both the breadth and the limits to variation within kind has come to be called baraminology. If we continue to use the term "species" for an interbreeding population, then baramin is a broader term that may include more than one species or reproductively isolated group. Consider, for example, the common fruit fly genus *Drosophila* (Greek for "fruit [*droso*] lover [*phila*]"). The species that is often used in high school and college genetics experiments is *D. melanogaster* (meaning "black [*melano*] bellied [*gaster*]"), but there are several other "sibling species" included in the same genus. They are called sibling because each species looks virtually identical to all the others, but they are given different species names because the "lookalike species" is reproductively isolated and cannot interbreed with any other. These different sibling fruit fly species, however, clearly descended from the same created kind and are members of the same baramin.

Are there objective, scientific tests that confirm that these various lookalike *Drosophila* species are all members of the same created kind or baramin? Yes. DNA pairing and chromosome matching in the lab show that these sibling species have the same genome, with sections of DNA moved to a new chromosome or a new position within the same chromosome and/or whole chromosomes split or joined to change chromosome number. Some species are reproductively isolated by slight differences in courtship ritual, resulting from either mutation or gene shuffling related to loci that affect behavioral traits—just allelic variation within kind, in either case. In short, the same criteria used to objectively and scientifically identify species can also be used to identify a broader category, either genus or baramin, when rearrangements of the same genetic information produce reproductively isolated subgroups of the same kind.

Using Linnaean terms with the sibling species of fruit flies, the genus name *Drosophila* represents the created kind and the lookalike species represent variation within kind. Alternatively, the sibling species could be called *fertilotypes* within the same baramin. Baramin-type scientific names would be somewhat similar or parallel to the genus-species names used by Linnaeus, but the baramin-type name established on objective, testable criteria would avoid the unscientific and flagrant subjectivity that is currently corrupting classification.

Reproductive isolation based on physical appearance and on ecology can also break up a created kind (baramin) into different sub-types. Fertilotypes look very similar but no longer interbreed, but *morphotypes* can still interbreed even though they look distinctly different from each other and usually live in different areas. Grizzly, black, and polar bears, for example, are all morphotypes within the same baramin. Their ability to interbreed when they come in contact establishes that they are indeed descendants of the same created bear kind. But they usually live in separate environments and their differences in appearance are distinctive and well-known—enough (though somewhat subjectively) to merit separate names, either baramin-morphotype, Linnaean genus-species, or species-subspecies. (The current assignment of bears to different genera is without scientific merit or objective defense.)

Morphotypes bring to visible expression the phenomenal variability God built into the first pair of each created kind. The bears that got off the Ark, for example, would be recognized as bears (like mongrel dogs are obviously dogs, though not from a distinct breed). As bears "multiplied and filled the earth" after the Flood, however, subgroups moved into different areas and carried different sets of alleles with them, developing the distinctive types now called grizzly, brown, black, sun, panda, and polar bears. Analogously, people that separated into different language groups at the Tower of Babel carried different proportions of alleles for skin color, height, and hair color, although all their descendants are obviously still human beings and can still intermarry/interbreed.

Environment can influence how the genetic makeup of an organism is expressed, especially among polygenic traits (those controlled by multiple genes). For instance, saplings among a stand of graceful weeping willow trees growing along a riverbank in a temperate zone will display a "stunted" dwarf phenotype if transplanted into an arctic region. Interestingly, a dwarf arctic willow can grow into a full weeping willow if transplanted to a temperate riverbank. The small Scottish red deer with very small antlers was about to be given a species name different from that of the larger European red deer—until Scottish red deer taken to New Zealand grew to a much larger size with large antlers. Fossils, the remains of pre-Flood created kinds, are often larger and may differ in other minor ways from their modern, post-Flood descendants. Many of these cases appear to be environmentally influenced polygenic-based variations of the same baramin.

The ability to classify subgroup species within distinct baramin is actually a problem for the evolutionary model. At the heart of evolution is the argument that all life forms have been generated by gradual change over eons of time from a common ancestor. However, if this was the case, should there be distinct, classifiable subgroup species at all? Even with the "missing links" still missing, should there be distinguishable species variations in a model that rests on slow and gradual change? Evolution would predict that organisms should blend together without distinct boundaries. Stephen Gould, a famed evolutionist, noted that biologists have been successful at identifying distinct and discrete species within the living world, and asked, "How could the existence of distinct species be justified by a theory [evolution] that proclaimed ceaseless change as the most fundamental fact of nature?"[6]

Without question, the ability to recognize and systematically classify distinct kinds and species is a strike against the evolutionary model of origins. Gould himself concludes that "the existence of distinct species was quite consistent with the creationist tenets of a pre-Darwinian era."[7] The clear division between kinds and the further classification of species variation supports the creation model of origins.

6. Gould, S. J. 1979. A Quahog Is a Quahog. *Natural History.* 88 (7): 18-26.

7. Ibid.

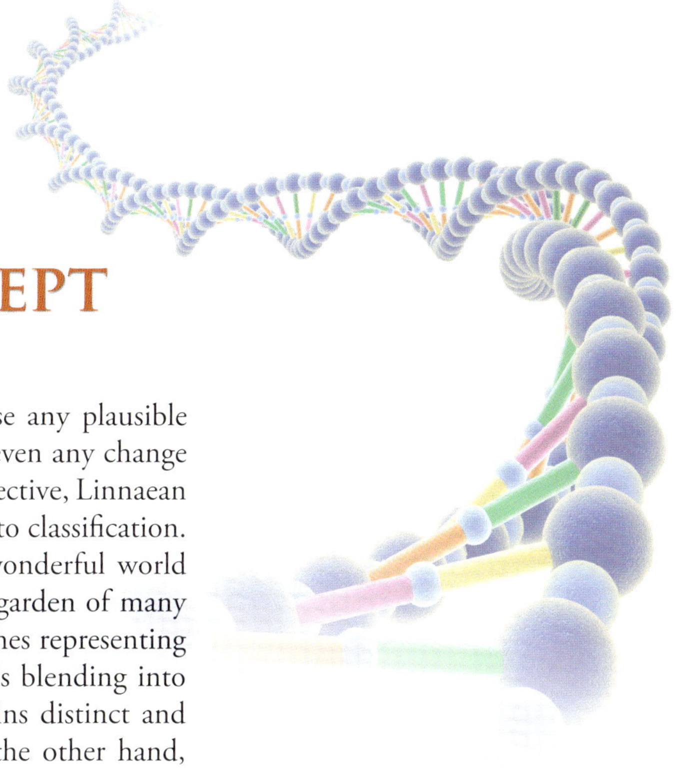

Diversity and the Mosaic Concept

Evolutionists have been completely unable to propose any plausible scientific explanation for the origin of species, or for even any change beyond variation within kind. Their gross misuse of objective, Linnaean (creationist) scientific naming has introduced chaos into classification. Present science verifies what the Bible records: the wonderful world of living things is like an orchard or gorgeous global garden of many beautiful bushes, each bush a created kind with branches representing subgroups, the tremendous diversity of different kinds blending into a harmonious whole, though each bush (kind) remains distinct and separate from all the others. Darwin's followers, on the other hand, boldly assert that evolutionary ties among organisms show branching patterns of descent in one single "tree of life" that "proves" that the great diversity of living things all evolved from a common ancestor.

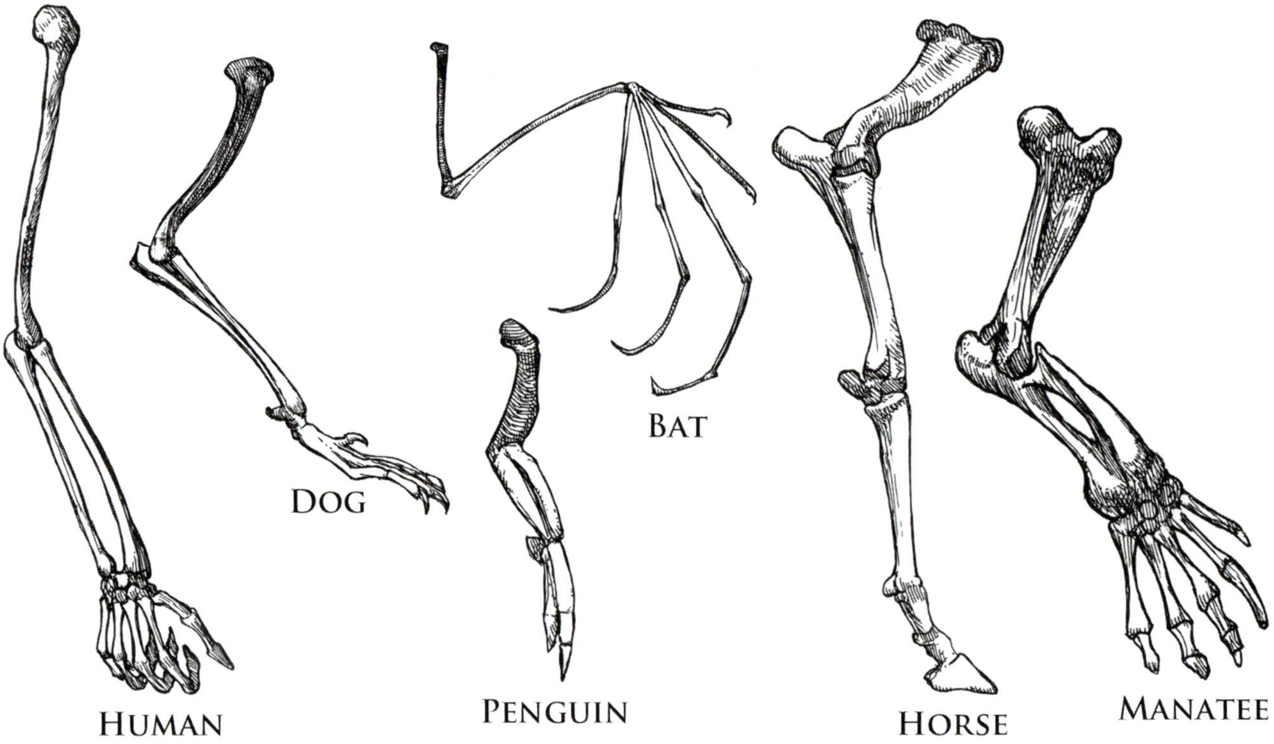

Figure 2: Human arm and animal forelimb bone structures

Homology: 1: a similarity often attributable to common origin. 2: a likeness in structure between parts of different organisms (as the wing of a bat and the human arm) due to evolutionary differentiation from a corresponding part in a common ancestor[8]

In an attempt to argue for common ancestry, evolutionists appeal to biological homology, noting the similarities in structures between animals and humans. For example, consider the human arm pictured in Figure 2. The human arm is constructed from numerous bones. The large upper bone is attached to the shoulder. The forearm contains two bones. The hand contains many bones in the wrist and bones that extend out into the fingers. It is evident from the figure that various animals have forelimbs that show a similar pattern.

Why is a human arm so similar (homologous) in structure to the limbs of various kinds of animals? Does this show or "prove" evolutionary descent? Evolutionists claim that the similarities, or the homology, between different life forms is evidence of evolution. They insist that common features between animals and humans point to descent from a common ancestor. In fact, Darwin himself said, "The similar framework of bones in the hand of a man, wing of a bat, fin of the

8. Homology. *Merriam-Webster Online Dictionary.* 2010. Merriam-Webster Online. Posted on merriam-webster.com/dictionary. Note that this definition presumes evolution to be true.

porpoise, and leg of the horse…and innumerable other such facts, at once explain themselves on the theory of descent with slow and slight successive modification."[9]

However, is there another interpretation of the data that could also explain the similarities in structure between animals and humans? As opposed to the evolutionary model of origins, the creation model of origins interprets data from a belief that a transcendent Creator God created all animals and humans as detailed in Genesis. Therefore, the biblical record of a supernatural creation provides a logical alternative explanation of the similarities (homology) in body parts and their functions between animals and humans. Instead of descent from a common ancestor, the homology of structures can be attributed to creation according to an efficient design that re-uses basic biological structures to accomplish similar purposes in the different created kinds. Is it more logical to assume that similar (homologous) structures seen between animals and humans are due to evolution over millions of years from a common ancestor—or that the Creator God created distinct kinds in the animal kingdom and humans using similar design features, while also providing optimal biomechanical design through variations and modifications?

The argument for homology as evidence for evolution from a common ancestor has its own limitations. As noted above, many similarities of the forelimbs exist between animals and humans. However, homology *within* an animal is problematic to the evolutionary model. For example, there are distinct similarities in structure between the forelimbs and the hind limbs of animals. Should it be concluded that hind limbs evolved from forelimbs, or that both evolved from an animal with a single limb whose descendants somehow developed additional homologous appendages?

Another strike against homology as proof of common ancestry is the recognizable correspondence between the male and female reproductive systems. Descent from a common ancestor would require that males evolved from females, females evolved from males, or both evolved from animals that did not have separate sexes. Evolutionists cannot explain the coordinated interaction between male and female reproductive systems in animals and humans.

A bigger obstacle for evolution is the existence of homologous structures that clearly have developed from non-homologous genetic information. In other words, very different genetic information and cellular mechanisms can produce homologous structures, which goes against the idea that similar structures result from genes that were modified during evolutionary descent. For example, both frogs and humans have five digits (fingers/toes) on each limb. However, these

9. Darwin, C. 1872. *The Origin of Species By Means of Natural Selection,* 6th ed. London: John Murray, 420.

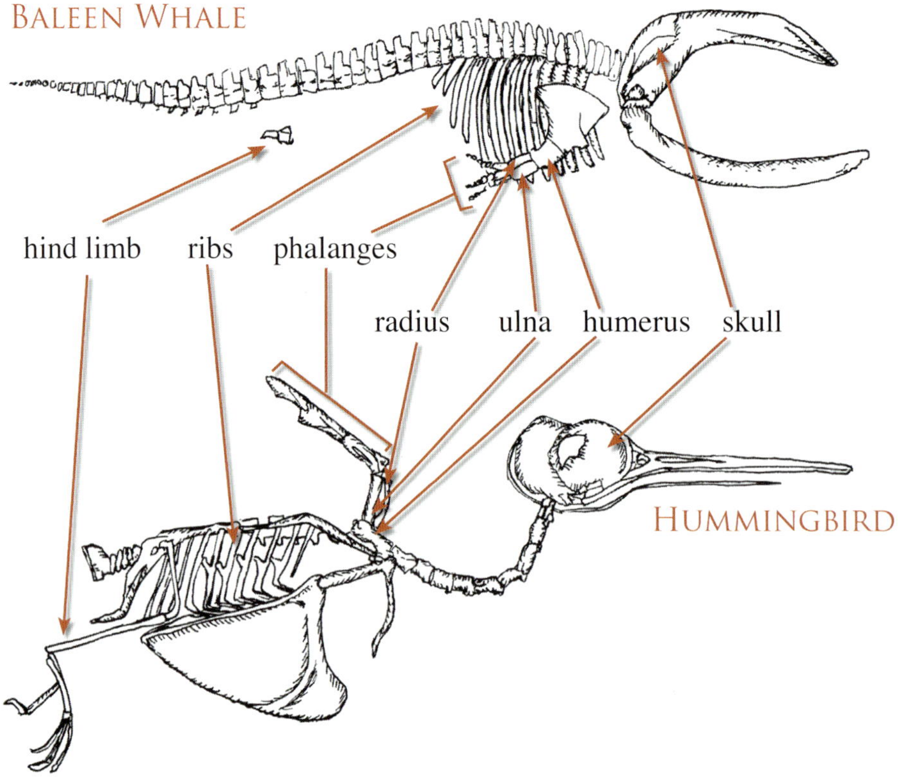

Corresponding bones between two different creatures are considered "homologous" only if those creatures are assumed to have evolved from a common ancestor. (Whale and hummingbird not to scale.)

digits form by very different mechanisms. On frogs, the digits grow out from buds. In contrast, fingers form in human embryos as the tissue between them is reabsorbed.

Many other examples exist that clearly show that homologous structures can result from very different mechanisms, indicating non-homology at the genetic and cellular levels. There is homology both between animal kinds, and between animals and humans. In his book *Evolution: A Theory in Crisis*, Michael Denton notes that this similarity exists "whether the causal mechanism was Darwinian, Lamarckian, vitalistic, or even creationist."[10] Given the alternatives, the current knowledge of homology strongly favors the creation model—the repeated use of efficient designs utilizing common biophysical concepts.

Convergence: independent development of similar characters (as of bodily structure of unrelated organisms or cultural traits) often associated with similarity of habits or environment[11]

An even stronger case for the creation model in the realm of anatomical similarity is *convergence*. The similarity of some anatomical

10. Denton, M. J. 1985. *Evolution: A Theory in Crisis.* London: Burnett Books, 155.

11. Convergence. *Merriam-Webster Online Dictionary.* 2010. Merriam-Webster Online. Posted on merriam-webster.com/dictionary.

structures cannot be explained by the evolutionary model. Therefore, this homology is classified as "convergent evolution." In this case, evolutionists claim that the similar structures resulted independently by descent from *different* ancestors. The classic example of convergence is the similarity between the eyes of vertebrates and certain cephalopods (squid and octopuses). The similarity between the visual organs of these very dissimilar taxa is obvious. For instance, the cornea, lens, and retina display the same configuration, detailed design, function, and biochemistry. Evolutionists struggle with proposing a common ancestor with traits that explain these similarities. Convergence, in fact, strongly favors the creation model, since evolutionary descent from a common ancestor cannot be used to explain the similarity of these complex anatomical structures—which are found fully formed in their respective creatures from their earliest appearance in the fossil record.

Probably the most quoted "data" in support of evolution is the claim that DNA comparisons show that there is 98 percent similarity between man and chimpanzee. The scientific community hopes that similarities in DNA sequences ("molecular homology") can provide a new basis for evolution-based classification. When ape and human phenotypes are compared, it is obvious that apes and man have many similarities. However, many significant physical differences also exist. Since apes and man exhibit many functional similarities in physical structure, it should not be surprising that DNA similarities exist—especially in the genes that govern basic cellular functions.

The report that humans and chimps have a 98 percent DNA homology is typically presented as an indication of common ancestry. This similarity, however, hardly tells the whole story. It is not just the parts involved that matter, it is also how they are assembled and the ways in which they work together. A 2006 *Discover* magazine article based upon the supposed 2 percent difference in DNA sequences presented that difference as favorable evidence for evolution. The author, Robert Sapolsky, however, admitted that "regulation is everything," a fact that has been emphasized by many creation scientists. He pointed out that the brains of man and chimp operate using "the same basic building blocks" that achieve "vastly different outcomes." According to Sapolsky, "there's not the tiniest bit of scientific evidence that chimps have aesthetics, spirituality, or a capacity for irony or poignancy."[12] He stated that the vast difference between the chimp and human brain's

12. Sapolsky, R. The 2% Difference. *Discover*. Published on discovermagazine.com April 4, 2006.

functional output or capacity can be credited to a "relatively few" genes that regulate brain cell production. But the number of cells that are produced, whether by these or other genes, is not the most operative factor. Many humans have dysfunctional or diseased brains that are the same size as normal brains. What makes the biggest difference between brains is not their sizes, but gene regulation, how the cells are connected, and ultimately how the brain functions. Gene regulation involves intracellular communication networks that determine what time, how much, how long, and which version of a gene is to be produced in a cell at any given moment.

It is important to understand that what defines a living thing (and its functioning parts) is not the substances from which it is derived, but the organization and resulting functionality of those substances. From a materials standpoint, all living things are essentially 100 percent identical. In other words, everything is ultimately composed of the same "materials" at the atomic level. However, the organization and regulation of these materials is the issue. Even if humans and chimps showed 98 percent DNA similarity, the 2 percent of DNA that differed could tremendously impact organization and regulation, yielding vast differences in phenotype. The *Discover* article mentioned above even states that "a tiny 2% difference translates into tens of millions of AGCT [DNA base] differences." Consider the potential differences even if humans and chimps actually are 98 percent homologous (which does not appear to be the case). A genome size of three billion base pairs would result in about 60 million base pair differences between man and ape. Sapolsky admitted that "there are likely to be nucleotide differences in every single gene."[13]

Even though gene function/regulation is more critical to consider than sequence similarity, it is important to take a closer look at the 98 percent similarity claims made by evolutionists. The first obvious contradiction to this claim of a 2 percent DNA difference is a 21 percent difference in overall genome size based on currently assembled contiguous DNA sequence between man and chimp.[14] In addition to the large genome size difference, the figures given in the 98 to 99 percent similarity range were taken from preselected coding regions. They were hand-picked for comparison based on the fact that they were already known to be similar. The dissimilar but critical regulatory sequences within and around genes are often not included in a comparative DNA analysis. Such sequences typically do not align well and reveal gaps. Not only are these key sequences not convenient to work with, but their inclusion also gives much lower DNA similarity estimates, which is bad for evolution because they highlight the many critical differences, especially between the human and chimp genomes.

A recent study compared segments of chimp chromosome 22 with

13. Ibid.

14. Current sequencing and assembly statistics for the human genome project and the chimpanzee genome project can be found on the UCSC Genome Bioinformatics website at genome.ucsc.edu.

homologous counterparts on human chromosome 21.[15] Because the study did not take into account DNA sequence differences due to large blocks of insertions and deletions—of which there were about 68,000 ranging in size from just a few bases to more than 450 bases—it is difficult to say what a truthful and accurate percentage difference in DNA similarity would be. The study reported a 1.4 percent difference between human and chimp based on only single base changes, but the real difference would obviously be much greater if all the data were included. The researchers did indicate an 83 percent protein sequence similarity between coding DNA for over 250 different proteins. This was a much lower level of similarity than anticipated.

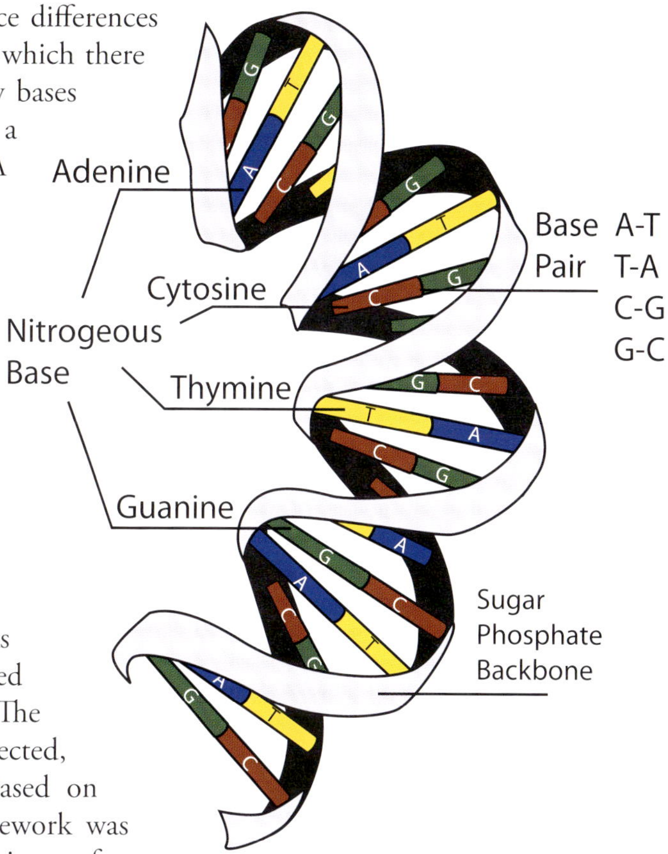

Besides all this, the chimp genome sequence is not an independent, stand-alone, finished product; it is only a low-resolution 6X-coverage rough draft containing thousands of un-oriented random blocks of sequence. So how was this mess initially sorted out? By using the human genome as a template! The DNA sequence "reads" were all oriented, connected, and assigned to specific *human* chromosomes. Based on evolutionary reasoning, the human genomic framework was chosen as a guide to assemble the thousands of pieces of un-oriented chimp sequence. As a result, we really do not know what the overall similarity is between the genomes because there has never been an unbiased DNA sequence analysis performed to give an accurate figure. Nor is the chimp genome sequence as complete as the human sequence, or assembled in an independent or unbiased way. But even so, it is obvious that the human-chimp similarity is much lower than 98 percent. There are hundreds of genes and genomic regions found in humans but not in chimps that have not been seriously studied and published, but alluded to "off the record" by a variety of key scientists in the genomics community.

Evolutionists have attempted to use molecular homology between various organisms to determine evolutionary ages. By analyzing the differences in the genetic code between two different but presumed related creatures, it was assumed that it could be determined when those species "diverged." Molecular clocks are controversial, however, because they are subjective and problematic in their development and application. One problem is that some proteins or genes may be very similar between organisms, while others are very different. So which genes or proteins should be chosen to compare? To make matters worse, some genes in one organism are not present in the one to which it is being compared.

15. The International Chimpanzee Chromosome 22 Consortium. 2004. DNA sequence and comparative analysis of chimpanzee chromosome 22. *Nature.* 429 (6990): 382-388.

The other major issue that comes into play is the use of supposed fossil-derived timeframes and evolutionary models to "calibrate" the molecular clock. This often involves frequent disagreements among evolutionists regarding timeframes and lineages, which adds further ambiguity. Because of all these factors, it is doubtful that any consensus will be reached about evolutionary molecular clocks any time soon in the scientific community. The available molecular data, however, fit the creation model quite well.

Instead of presuming evolutionary time at the outset, one can apply reasonable mutation rates to the DNA sequences being compared in order to extrapolate an elapsed time. This most often provides an estimate that is more compatible with known biblical ages than with evolution's contrived ages. For example, there are merely 22 out of 16,569 DNA differences between any living woman and the most likely mitochondrial DNA sequence for Eve. After only a few mutations per generation—as is consistent with observations—Eve would have been alive about 6,000 years ago, according to this data.

However, when either a creationist or an evolutionist tries to use these molecular clocks, they both run into a big problem—mutation rates are not very reliable. Mutations accumulate much faster in certain places in the genome, called mutational hotspots. In addition, DNA repair mechanisms can reverse some mutations. Therefore, a more reliable method for estimating dates is needed, and God's Word provides that.

Michael Denton presents the misfit of molecular data in light of two competing evolutionary views. In summary, he states that:

> The difficulties associated with attempting to explain how a family of homologous proteins could have evolved at constant rates [have] created chaos in evolutionary thought. *The evolutionary community has divided into two camps—those* still adhering to the *selectionist* position, and those rejecting it in favour of the *neutralist.* The devastating aspect of this controversy is that neither side can adequately account for the constancy of the rate of molecular evolution, yet *each side fatally weakens the other.* The selectionists wound the neutralists' position by pointing to the disparity in the rates of mutation per unit time, while the neutralists destroy the selectionist position by showing how ludicrous it is to believe that selection would have caused equal rates of divergence in "junk" proteins or along phylogenetic [presumed evolutionary] lines so dissimilar as those of man and carp. Both sides win valid points, but in the process the credibility of the molecular clock hypothesis is severely strained and with it the *whole paradigm* [*worldview*] *of evolution itself is endangered.*[16]

16. Denton, 305, emphasis added.

Denton not only offers these devastating remarks against the evolutionary position, but also describes molecular homology data as a "biochemical echo of typology." Why is this significant? Because typology is the pre-Darwinian view of classification developed by scientists on the basis of creationist thinking.

Protein sequences tell the same story. Evolutionists Richard E. Dickerson and Irving Geis attempted to show how molecular homology could be used to support evolutionary lines of descent. In doing so, they were forced to acknowledge that hemoglobin presented "a puzzling problem." Hemoglobin is a protein that carries oxygen in red blood cells. However, Dickerson and Geis noted that "hemoglobins occur sporadically among the invertebrate phyla [creatures without a backbone] in no obvious pattern."[17] In other words, hemoglobins do not show the expected evolutionary branching pattern. Instead, they show what creationists would predict—a *mosaic* pattern.

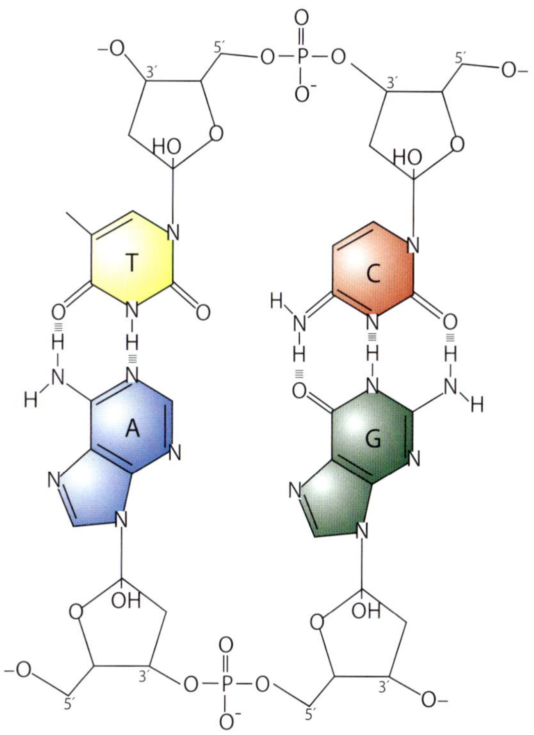

AGCT (DNA Base)

Hemoglobin is found in nearly all vertebrates. But hemoglobin is not just limited to vertebrates. It is also found in among certain groups of earthworms, starfish, clams, insects, and even in some bacteria. The hemoglobin protein is complete and fully functional in all these vastly different organisms. How does this impact the evolutionary model of common descent? Dickerson concludes, "It is hard to see a common line of descent snaking in so unsystematic a way through so many different phyla."[18]

If the evolutionary model of origins is correct, hemoglobin's evolution should be traceable. However, this is not the case. Dickerson raised the possibility that hemoglobin could potentially serve as an example of repeated evolution, the spontaneous appearance of hemoglobin in numerous groups of organisms independently. Dickerson, though, contends that convergent evolution could really only be accepted as a possibility if just the final structure and function of hemoglobin were under consideration. However, as we have seen throughout this study, many factors contribute to phenotypic structure and function. Therefore, Dickerson concludes that it does not seem possible that the entire eight-helix folded pattern of hemoglobin appeared repeatedly by time and chance. The creation model has a logical explanation for the appearance of complete, fully functional hemoglobin in a vast array of organisms—it was placed there by the Creator's design as He deemed necessary to complete the particular pattern of the life form He was creating.

17. Dickerson, R. E. and I. Geis. 1969. *The Structure and Action of Proteins.* New York: Harper and Row.

18. Ibid.

God used different genes and gene families over and over again in different combinations and proportions to make a variety of life forms, somewhat like an artist might use several different kinds of colored stones over and over in different proportions and arrangements to make a variety of artistic designs.

Mosaic: 1: a surface decoration made by inlaying small pieces of variously colored material to form pictures or patterns; 3: something resembling a mosaic[19]

As opposed to the evolutionary model's branching pattern, the creation model can be likened to a mosaic pattern. Different kinds of colored stones can be used by an artist to assemble an untold number of artistic designs, depending on the proportions and arrangements of the individual stones. The reds, greens, blues, browns, and yellows that form a geometric pattern in one picture may be used to depict a grouping of flowers in another. It is logical that in the same fashion God has repeatedly utilized a number of different genes and gene families that, depending on their combinations, proportions, and regulation, bring forth the vast array of living organisms that exist today. Each created kind is a distinct combination of various traits. Most likely you have seen other people with your eye color, your type of hair, your nose, your height, or your face shape. But you are uniquely you because no one else has your combination of traits, even if your individual "parts" are similar to others.

19. Mosaic. *Merriam-Webster Online Dictionary.* 2010. Merriam-Webster Online. Posted on merriam-webster.com/dictionary.

The mosaic concept also applies to the distribution of traits among the various different kinds of life. Darwin's followers had hoped to find series of incomplete structures showing gradual evolution of different features along branching lines of descent from "primitive" common ancestors to more "advanced" descendants. Scientists, however, found exactly the opposite. Whether it is a single molecule like hemoglobin or a complex structure like a wing, features are just as complete and complex ("advanced") from their appearance in earlier life forms as they are among modern life forms. Furthermore, these traits, like colored stones in an artist's mosaic, are distributed according to a plan and purpose in the creative artist's mind, not at all along branching lines of evolutionary descent.

Evolutionists face practical problems in dealing with the mosaic pattern of trait distribution. This can be demonstrated by the classification of algae. Typically, the trait by which algae are classified into major groups is their pigment color. But if they are grouped this way, it means that their structural complexity and type of sexuality must have evolved independently on separate branches of the supposed evolutionary true. Alternatively, algae can first be classified based on their level of structural complexity. This approach, however, also has issues because then neither the color pattern nor the type of sexuality can be traced back to a common ancestor. Using the type of sexuality as the starting category also does not solve the problem. In this case, the resulting evolutionary tree contradicts the trees based on pigment and structural complexity. The bottom line is that the scientific evidence is in opposition to any attempted construction of an evolutionary history for these organisms.

When it came to the very large, well-preserved phylum of shelled bivalves called brachiopods or lampshells, the author of a classic textbook on invertebrate fossils said that the classification of these shelled creatures was "impossible." Then he proceeded to classify these "brachs" in great detail into many precisely and objectively defined groups (a memory nightmare for generations of students to come!). How did he do the impossible? *He switched from evolution to science.* When he called the classification of lampshells "impossible," he meant that an evolutionary classification based on branching lines of descent was impossible, producing numerous contradictory "trees" with the supposed evolution of one trait clashing with beliefs about traits based on other "trees." Then, in effect, the textbook writer gave up his chaotic and fruitless search for imaginary common ancestors and mythical evolutionary trees and looked instead, like a creationist would, at the numerous detailed features of well-preserved fossil shells collected from all over the world. When he switched from trying to classify things based on non-scientific evolutionary beliefs to looking at scientifically observable traits, the author developed a marvelous system of classification still in use today—a system, whether consciously or not, that is based on Linnaean (creationist) principles!

The mosaic concept both for defining created kind and for describing trait distribution among a diversity of life forms might also be called the modular concept. *Mosaic* emphasizes the plan and purpose in the Creator's mind; *modular* emphasizes the use of "common stock building materials" a contractor might use for many different projects—e.g., bricks for building a house, fence, fireplace, street, or office complex. It seems that God used "common stock building materials," like bones, to produce arms, legs, flippers, and wings, both as evidence of theme and variation, and also to show that there is only one Creator God, not many, and one plan for creation working toward its consummation in the coming again of Jesus Christ.

Interestingly, when we go to the genome, the molecular basis for observed homologies in anatomy, we see a similar modular trend, just as we see in a software programmer writing code. To effectively develop large software programs, code is written in a modular and integrated format. Large-scale studies of gene function for many traits are confirming that genes are expressed in integrated groups of specific modules (sets of genes). In fact, most of what makes humans biologically unique when compared to chimps and other animals is not so much sequence similarity, but how modules of genes are regulated in the genome and their interaction with other modules. Recent studies are demonstrating clear differences in gene network expression patterns between humans and chimps, and the largest differences are observed in regard to brain function, dexterity, speech, and other traits with cognitive components.

What is most remarkable about the human genome is that it contains about 25,000 protein-coding genes, but more than 1,000,000 protein variants have been discovered, a marvel of incredible regulatory complexity. We now know that genes contain multiple transcription start and stop sites, signal sequences for alternatively splicing multiple sections of the resulting RNA transcript in various patterns, and non-protein coding regions in and around the gene that code for small regulatory RNAs that influence gene regulation. Also, multiple regulatory sequences that interact with various regulatory proteins exist not only directly upstream from the gene, but downstream, within it, and as much as one million base pairs away. Mutations in regulatory sequences can be more harmful than those in coding regions, since they can cause a gene to shut down or go wildly out of control.

Even at the individual gene level, the modular approach holds true as each gene is actually involved in a complex grouping of regulatory and coding information that is re-used and manipulated in many cases to produce a wide variety of protein variants in response to a wide variety of mechanisms. In fact, the federally funded ENCODE project uncovered transcriptional activity in greater than 90 percent of their

20. The ENCODE Project Consortium. 2007. Identification and analysis of functional elements in 1% of the human genome by the ENCODE pilot project. *Nature*. 447 (7146): 799-816.

studied portion of the human genome.[20] It was once thought that roughly 95 percent of the genome was just "junk DNA," but creation scientists correctly predicted that design functions would be discovered for this DNA.

The mosaic/modular concept might also be called the matrix concept. *Matrix* would refer to mathematical precision and even predictability in God's plan. Beginning with Russian chemist Dmitri Mendeleev, chemists have used the flat (two-dimensional, or 2D) periodic table—a chemical mosaic—to arrange the fewer than 100 known elements used to make millions of chemical compounds. "Periodic" refers to repeated patterns, like the magazines or "periodicals" that come out once a month or so. Branching evolutionary trees were never even tried by chemists (and do not work for biologists), but the 2D grid (periodic table or chemical mosaic) reflected a pattern Mendeleev used to predict, and characterize in advance, new elements that were then unknown to science.

The platypus is an extraordinary example of a mosaic combination of traits. It has mammal fur, chicken spurs, a duck bill, and webbed feet, and also produces its offspring through reptile-like eggs.

The *mosaic/modular/matrix concept* may serve biology in two different but interrelated ways. As discussed earlier, it is logical to believe that God used a number of genes and gene sets as the essential building blocks for the vast array of life. The arrangement, organization, number, and kind of these non-unique genetic building blocks have provided an objective "formula" by which the various unique living organisms (each within its own created kind) can be identified. In addition, created kinds can be further differentiated based on the diversity of traits within the different groups of created kinds. This allows them to be classified in various distinct categories. God has evidently created many different kinds of life, each with a more or less limited but always important role to play in His total plan for the global ecosystem or biosphere. Each kind would then be analogous to a member of the church, the body of Christ, each with unique gifts designed to benefit, and to be benefitted by, all the others.

When Mendeleev discovered the pattern God used in creating the chemical elements, he was able to predict the existence and properties of elements not then known to science. Using a multi-dimensional mathematical matrix to display distinctions of each presently known created kind, creationists may one day discover predictive patterns of trait distribution among living things, enabling them to find and even describe ahead of time organisms not yet known to science, and prediction is the real measure of merit among scientific theories.

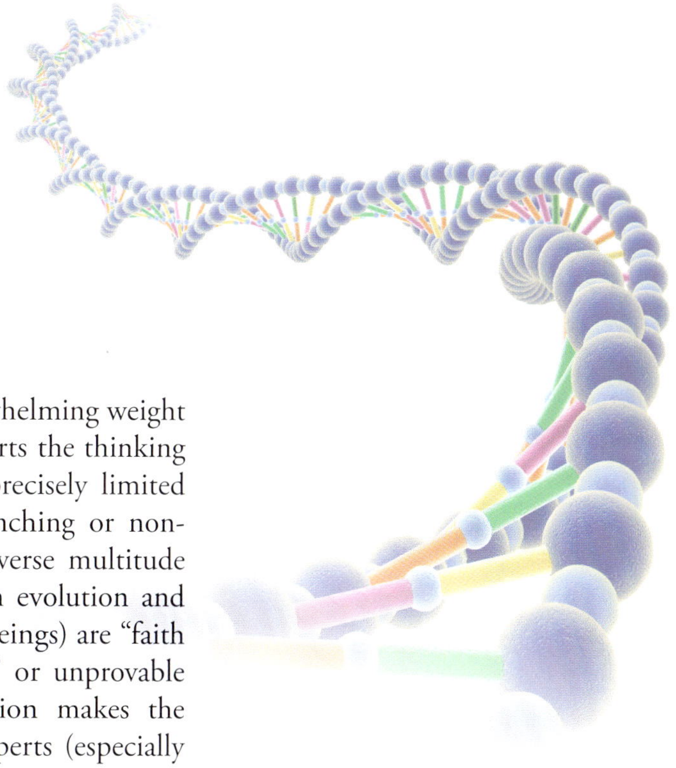

Conclusion

When it comes to the kinds of life we study, the overwhelming weight of scientific evidence observable today strongly supports the thinking of creation scientists regarding both 1) broad but precisely limited variation within kind, and 2) the mosaic (non-branching or non-evolutionary) distribution of traits among earth's diverse multitude of different created kinds. At bottom, however, both evolution and creation (like every other enterprise of finite human beings) are "faith based." Both are built on "religious presuppositions" or unprovable assumptions derived from one's worldview. Evolution makes the faith-based assumption that the words of human experts (especially Darwin) are the surest guide to understanding our past, present, and future. Creationists make the faith-based assumption that God's Word, not man's, is the surest guide to understanding God's world, from Generation through Degeneration to Regeneration (Creation, Corruption, Catastrophe, Christ).

So what is the difference? Evolution is a faith that the observable facts have failed. Wanting to believe experts in science, Darwin's followers are dismayed to find their beliefs at odds with the scientific evidence. Like Darwin, they face the choice of throwing out the scientific facts or throwing out present evolutionary concepts, either to build a new view of evolution, or to build an entirely new origins model.

Creation, on the other hand, involves a faith that fits the facts, which builds both confidence in God's Word and enthusiasm for deeper scientific study of God's world. Former evolutionists are invited to share that worldview and its view of the new world to come—the world where life wins: rich, abundant, eternal new life in Christ, our Creator and Redeemer!

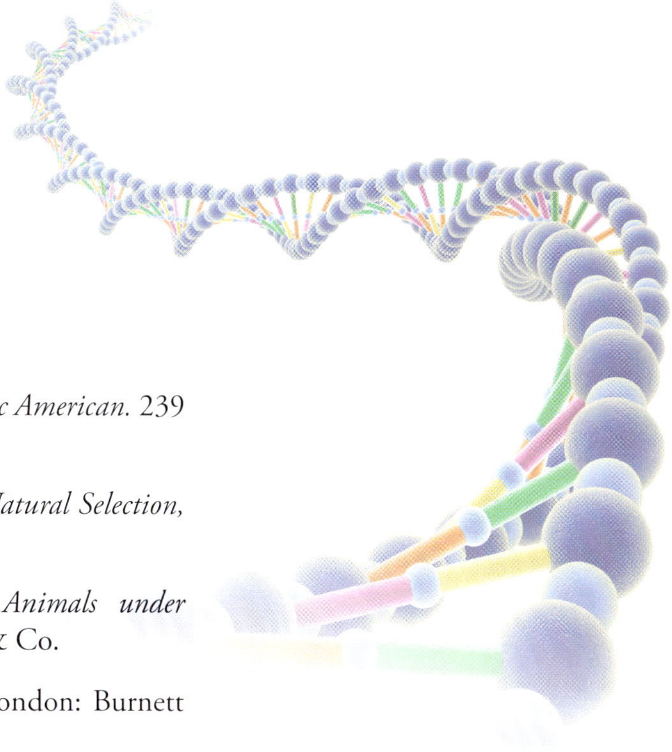

BIBLIOGRAPHY

Ayala, F. 1978. The Mechanisms of Evolution. *Scientific American.* 239 (3): 56-69.

Darwin, C. 1872. *The Origin of Species By Means of Natural Selection,* 6th ed. London: John Murray.

Darwin, C. 1883. *The Variation of Plants and Animals under Domestication*, 2nd ed., rev. New York: Appleton & Co.

Denton, M. J. 1985. *Evolution: A Theory in Crisis*. London: Burnett Books.

Dickerson, R. E. and I. Geis. 1969. *The Structure and Action of Proteins.* New York: Harper and Row.

Evolution: A Scientific American Reader. 2006. Chicago: University of Chicago Press.

Gould, S. J. 1979. A Quahog is a Quahog. *Natural History*. 88 (7): 18-26; also published in Gould, S. J. 1979. Species Are Not Specious. New Scientist. 83 (1166): 374-376.

Knaub, C. W. 1983. *A Critique of Molecular Homology.* Santee, CA: Institute for Creation Research (master's thesis).

Parker, G. 2007. *Building Blocks in Science.* Green Forest, AR: Master Books.

Parker, G. 2006. *Creation: Facts of Life: How Real Science Reveals the Hand of God*. Green Forest, AR: Master Books.

Sapolsky, R. The 2% Difference. *Discover*. Published on discovermagazine.com April 4, 2006.

Image Credit

Susan Windsor: 12, 20, 28, 31, 33

For More Information

Sign up for ICR's FREE publications!

Our monthly *Acts & Facts* magazine offers fascinating articles and current information on creation, evolution, and more. Our quarterly *Days of Praise* booklet provides daily devotionals—real biblical "meat"—to strengthen and encourage the Christian witness.

To subscribe, call 800.337.0375 or mail your address information to the address below. Or sign up online at www.icr.org.

Visit ICR online

ICR.org offers a wealth of resources and information on scientific creationism and biblical worldview issues.

- ✓ Read our daily news postings on today's hottest science topics
- ✓ Explore the Evidence for Creation
- ✓ Investigate our graduate and professional education programs
- ✓ Dive into our archive of 40 years of scientific articles
- ✓ Listen to ICR radio programs
- ✓ Order creation science materials online
- ✓ And more!

For more information, contact:

INSTITUTE *for* CREATION RESEARCH

P. O. Box 59029
Dallas, TX 75229
800.337.0375

SCIENCE EDUCATION ESSENTIALS

How do I explain the differences between biblical creation and evolution?

What evidence for the origin of life should my students know?

Where do I go for trustworthy information on science research and education?

For 40 years, the Institute for Creation Research has equipped teachers with evidence of the accuracy and authority of Scripture. In keeping with this mission, ICR presents Science Education Essentials, a series of science teaching supplements that exemplifies what ICR does best—providing solid answers for the tough questions teachers face about science and origins.

This series promotes a biblical worldview by presenting conceptual knowledge and comprehension of the science that supports creation. The supplements help teachers approach the content and Bible with ease and with the authority needed to help their students build a defense for Genesis 1-11.

Each teaching supplement includes a content book and a CD-ROM packed with K-12 reproducible classroom activities and PowerPoint presentations. Science Education Essentials are designed to work within your school's existing science curriculum, with an uncompromising foundation of creation-based science instruction.

Demand the Evidence. Get it @ ICR.

Origin of Life

How did life get started on earth? Many scientists believe that life began from natural processes, but the Bible presents an alternate explanation.

Origin of Life, the first of the series, answers basic life questions such as:

- What is the origin of life?
- What are the physical and biblical definitions of life?
- What are the physical requirements for life?
- Can life exist elsewhere in the solar system?
- And much more

It gives scientific explanations for the chemical basis for life from a biblical worldview and discusses the efforts to create life in the laboratory. Most importantly, it offers scientific evidence proving that the creation of life "requires an act of God."

Available for **$24.95** (plus shipping and handling)

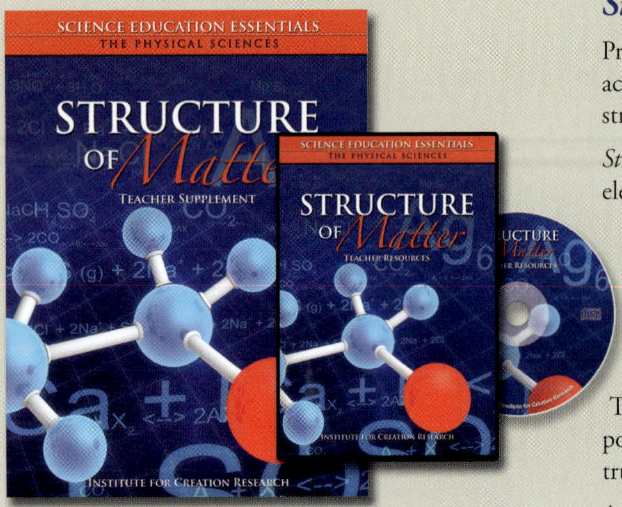

Structure of Matter

Predictions in science are based on knowledge of observable events. The accuracy with which science can make predictions points to the order and structure God established within His created universe.

Structure of Matter, the second of the series, explores structural forces and elements of nature such as:

- The First and Second Laws of Thermodynamics
- The structure of the atom
- The periodic table
- Properties of matter
- And much more

The order and design of the universe point to a Creator of omnipotent power and omniscient strength. Truly, the structure of matter upholds the truth that "in the beginning God created the heaven and the earth."

Available for **$24.95** (plus shipping and handling)